Smart Home & IoT f Beginners:

A Practical Guide to Setting Up, Integrating, and Securing Your Home

E. Psaila

Smart Home & IoT for Beginners:

A Practical Guide to Setting Up, Integrating, and Securing Your Home

First Edition: **February 2025**

ISBN: 978-1-923432-37-6

Table of Contents

Chapter 1: Introduction: Welcome to the Smart Home Revolution

Welcome to the exciting world of smart homes and the Internet of Things (IoT)! This chapter serves as your gateway into a realm where everyday objects transform into intelligent devices that communicate with each other, making your life more convenient, efficient, and secure. Whether you're a complete beginner or just looking to demystify the jargon, this chapter will set the stage for everything you'll learn in this guide.

1.1 Overview & Motivation

The Promise of a Smart Home

Imagine waking up to a home that adjusts your thermostat before you even get out of bed, turns on your favorite morning playlist, and gradually brightens your lights as the sun rises. This isn't a scene from a futuristic movie—it's the potential of a smart home. At its core, smart home technology is about enhancing your daily life through automation and intelligent connectivity.

Key Benefits:

- **Convenience:**
  With a smart home, you can control lights, appliances, security systems, and more with a simple voice command or a tap on your smartphone.

No more fumbling for switches in the dark or wondering if you left the stove on.

- **Energy Efficiency:**
  Smart devices can help you monitor and reduce energy consumption. For instance, smart thermostats learn your schedule and adjust heating or cooling accordingly, which can lead to lower utility bills and a reduced carbon footprint.

- **Enhanced Security:**
  Integrated security systems, including smart locks, cameras, and sensors, offer real-time monitoring and alerts. This connectivity can provide peace of mind whether you're at home or away.

- **Personalization:**
  As these devices gather data on your habits and preferences, they can tailor your home environment to suit your lifestyle—from mood lighting and ambient music to automated routines that simplify everyday tasks.

The Evolution of IoT and Its Impact

The Internet of Things represents a significant shift in how we interact with technology. Over the past few decades, we've seen the transformation from isolated gadgets to interconnected devices that share data seamlessly over a network. Here's a brief look at how we got here:

- **Early Beginnings:**

  The concept of connecting devices dates back to simple automation systems in industrial settings. Over time, advancements in wireless technology and microprocessors paved the way for more sophisticated consumer devices.

- **Rapid Growth:**

  The last ten years have witnessed exponential growth in IoT devices. From smart speakers and wearable technology to interconnected appliances, these devices are now an integral part of modern living.

- **Future Trends:**

  As technology evolves, so too will our smart homes. Emerging trends like artificial intelligence, machine learning, and edge computing promise to make devices even smarter and more intuitive, ensuring that the future of home automation is both exciting and transformative.

Why This Guide?

Despite the undeniable advantages, many consumers feel overwhelmed by the technical details and industry-specific terminology. This guide is designed to break down these barriers by:

- **Simplifying Complex Concepts:**
  We translate technical jargon into everyday language, ensuring that even those with limited technical knowledge can understand and implement smart home solutions.

- **Providing Step-by-Step Instructions:**
  Each chapter is structured to offer detailed, actionable steps—from device selection and setup to troubleshooting and security measures.

- **Empowering You to Innovate:**
  With a solid foundation in smart home principles, you'll be equipped to not only use current technology but also adapt to and integrate future innovations.

1.2 What to Expect

A Roadmap for Your Smart Home Journey

This book is designed as a comprehensive guide that covers every aspect of building, integrating, and maintaining a smart home. Here's a sneak peek at what each chapter will offer:

1. **Introduction: Welcome to the Smart Home Revolution**
   This chapter lays the groundwork by explaining the benefits, history, and potential of smart home

technology.

2. **Understanding the Basics of IoT and Smart Homes**

We dive into what IoT is, how it works, and why it's revolutionizing the way we interact with our living spaces.

3. **The Building Blocks of a Smart Home**

Learn about the essential components, such as connectivity protocols and home networking, that power your smart devices.

4. **Choosing Your Smart Home Devices**

Explore different ecosystems, budget-friendly options, and key considerations for future-proofing your investments.

5. **Setting Up and Integrating Your Smart Devices**

Step-by-step guides help you get your devices up and running, from initial setup to integrating multiple platforms.

6. **Automating Your Home**

Discover how to create routines and schedules that make your home work for you, with practical examples for various scenarios.

7. **Troubleshooting Common Issues**

Detailed troubleshooting guides and tips to quickly resolve issues that might arise with your smart home devices.

8. **Securing Your Smart Home**
   Understand the risks and learn best practices to safeguard your devices and personal data from potential threats.

9. **Expanding and Future-Proofing Your Smart Home**
   Advice on scaling your smart home and integrating new technologies as they emerge, ensuring your system stays up-to-date.

10. **Hands-On Projects and DIY Integrations**
    Engage with practical projects that encourage you to experiment, customize, and enhance your smart home setup.

11. **Resources, Glossary, and Further Learning**
    A comprehensive resource section that includes a glossary of terms, additional reading, and links to online communities.

12. **Conclusion: Embracing the Future of Smart Living**
    A wrap-up that reinforces key lessons and inspires you to continue exploring the exciting possibilities of smart home technology.

Tailored for Beginners

This guide is specifically designed for beginners who might be intimidated by the technical nature of smart home technology. Here's how we ensure it remains accessible:

- **Clear, Concise Language:**
  We avoid overwhelming technical jargon by explaining terms in simple language and using relatable analogies.

- **Step-by-Step Instructions:**
  Every process is broken down into manageable steps, with detailed explanations and visuals where applicable, ensuring you can follow along even if you're new to technology.

- **Practical Examples:**
  Real-life scenarios and case studies illustrate how smart home devices can be integrated into daily life, making abstract concepts tangible and relevant.

- **Continuous Support:**
  Throughout the book, you'll find tips, checklists, and troubleshooting guides that serve as handy references for when you're setting up or expanding your smart home.

Final Thoughts

In this opening chapter, we've laid the foundation for understanding why smart homes are not just a luxury of the future but a practical reality today. We've explored the benefits that make smart technology a compelling choice for modern living and provided a roadmap for what you can expect in the chapters ahead.

As you embark on this journey, remember that every expert was once a beginner. This guide is here to empower you to transform your living space into an intelligent, responsive, and secure environment. The revolution in smart living has begun—let's dive in and start building your smart home, one step at a time.

Chapter 2: Understanding the Basics of IoT and Smart Homes

In this chapter, we dive into the core concepts behind the Internet of Things (IoT) and explore how these ideas lay the foundation for smart homes. Whether you're new to technology or simply curious about how everyday devices can communicate and work together, this chapter will break down the essential principles in clear, accessible language.

2.1 Defining IoT

What Is the Internet of Things?

The Internet of Things (IoT) refers to the network of physical objects—"things"—embedded with sensors, software, and other technologies to connect and exchange data with other devices and systems over the internet. These devices can range from everyday household items like refrigerators and light bulbs to more complex systems like security cameras and thermostats.

- **Key Characteristics:**

  - **Connectivity:**
    IoT devices are connected to the internet, which allows them to send and receive data.

- **Sensing:**
  Many devices are equipped with sensors that monitor environmental conditions (such as temperature, motion, or light) or even track user behaviors.

- **Data Processing:**
  Collected data is processed either locally or in the cloud, enabling the device to make intelligent decisions or provide valuable insights.

- **Automation:**
  Devices can operate autonomously, executing actions based on preset conditions or real-time inputs without human intervention.

How Do Devices Communicate and Share Data?

Understanding how IoT devices communicate is key to grasping the power of a smart home.

- **Communication Protocols:**
  IoT devices rely on various communication protocols that determine how they send and receive information. Some common protocols include:

  - **Wi-Fi:**
    Widely used for its high-speed internet connection, Wi-Fi is prevalent in home networks.

- **Bluetooth:**
  Ideal for short-range communication, Bluetooth is often used to connect devices like smart speakers and wearables.

- **Zigbee and Z-Wave:**
  These are low-power, mesh networking protocols designed specifically for home automation. They allow devices to communicate over longer distances by passing data through intermediary devices.

- **Data Exchange Mechanisms:**

  - **Cloud-Based Communication:**
    Many smart devices send data to a cloud server where it's processed and then returned to the device or another connected gadget.

  - **Local Communication:**
    Some devices communicate directly with each other on a local network, which can lead to faster response times and enhanced privacy.

- **Security Considerations:**
  Secure data exchange is crucial. IoT devices often use encryption and authentication protocols to protect data as it travels across networks.

Real-World Analogy

Imagine a busy post office where packages (data) are sorted, labeled, and sent off to their destinations. In this analogy:

- **Packages** represent the data being transferred.

- **Sorting machines and postal workers** symbolize the sensors and processors that analyze and forward data.

- **Delivery trucks** are like the communication protocols ensuring that data reaches its intended destination.

This system works around the clock, much like your smart home devices, constantly exchanging information to make your environment more efficient and responsive.

2.2 History & Trends

A Brief History of Smart Technology

The journey toward today's smart homes has been paved with innovation and technological milestones:

- **Early Automation:**
  The concept of automation dates back to industrial applications where machines performed repetitive tasks without human intervention. Early examples include assembly line robots and basic

programmable devices.

- **The Rise of Home Automation:**

  In the 1970s and 1980s, home automation began to emerge with devices like programmable thermostats and early security systems. However, these systems were typically expensive and required professional installation.

- **The Internet Revolution:**

  The expansion of the internet in the 1990s and early 2000s opened the door to more sophisticated connectivity. As broadband access improved, manufacturers began to integrate internet connectivity into consumer devices.

- **Modern IoT Era:**

  Over the past decade, advancements in wireless communication, miniaturization of electronics, and the development of user-friendly interfaces have democratized smart technology. Today, IoT devices are not only more affordable but also designed with the everyday user in mind.

Current Trends and Future Predictions

The smart home landscape is evolving rapidly. Here are some of the trends shaping its future:

- **Increased Interoperability:**
  Manufacturers are working towards standardizing communication protocols, which means that devices from different brands are more likely to work together seamlessly.

- **Enhanced AI Integration:**
  Artificial intelligence and machine learning are becoming integral to smart home systems. These technologies help devices learn from your habits, predict your needs, and automate processes more intelligently.

- **Focus on Energy Efficiency:**
  As sustainability becomes a global priority, smart homes are increasingly designed to optimize energy usage. This includes advanced climate control, smart lighting, and energy monitoring systems.

- **Improved Security Measures:**
  With the rise in IoT adoption, there is also an increased focus on cybersecurity. Future devices are expected to incorporate more robust security features to protect user data and privacy.

- **Edge Computing:**
  Instead of sending all data to the cloud, some devices are starting to process information locally. This not only speeds up response times but also

reduces reliance on internet connectivity and enhances data privacy.

Implications for the Consumer

Understanding the history and trends helps you appreciate why modern smart homes are both powerful and accessible. It also underlines the importance of selecting devices that are future-proof—ensuring they can adapt to evolving standards and technologies.

2.3 Everyday Examples

Enhancing Daily Life with IoT

Let's explore some real-world scenarios where IoT devices make a tangible difference in everyday life:

- **Smart Thermostats:**
  Imagine arriving home on a chilly winter day to a house that has already adjusted the temperature to your liking. Smart thermostats learn your schedule and adjust heating or cooling systems automatically, ensuring comfort and reducing energy costs.

- **Intelligent Lighting:**
  Smart lighting systems allow you to control brightness and color based on the time of day or your mood. For instance, dimming the lights for a relaxing evening or brightening them for focused work is just a voice command or smartphone tap

away.

- **Security and Surveillance:**
  Smart cameras, doorbells, and locks provide enhanced security. With features like motion detection, real-time alerts, and remote access, you can monitor your home from anywhere, ensuring peace of mind even when you're away.

- **Voice Assistants:**
  Devices such as Amazon Alexa, Google Assistant, or Apple Siri integrate with various smart devices, enabling hands-free control of your home. You can ask your assistant to play music, provide weather updates, or even control your lighting system.

- **Smart Appliances:**
  Modern refrigerators can notify you when you're running low on groceries, while washing machines can be scheduled to run during off-peak hours to save on energy. These appliances make routine tasks more efficient and less time-consuming.

Bringing It All Together

Consider a typical morning in a smart home:

1. **Wake-Up Routine:**
   Your smart alarm clock gently wakes you up by gradually increasing the light in your room. As you stretch, the thermostat adjusts the room temperature, and the smart speaker begins playing

your favorite morning playlist.

2. **Kitchen Efficiency:**

   In the kitchen, your smart coffee maker starts brewing your coffee just as you enter. The refrigerator displays a list of items that need to be restocked, and you receive a notification on your smartphone.

3. **Secure Departure:**

   As you leave for work, you lock the doors remotely, and the security system arms itself. You check your home's status on your phone, confident that everything is secure and running as scheduled.

Why These Examples Matter

By illustrating how IoT devices function in real-world scenarios, you can see that smart home technology isn't just about gadgets—it's about enhancing the quality of life. These everyday applications show how connectivity and automation simplify tasks, improve efficiency, and even contribute to energy savings and security.

Final Thoughts

In Chapter 2, we've built a solid foundation by defining what IoT is, exploring its historical evolution, and examining current trends. We've also seen practical examples of how smart technology can transform everyday routines. With this understanding, you're now better equipped to appreciate the intricacies of smart homes and ready to delve deeper into the technical components in the coming chapters.

As you move forward, remember that each smart device and system is a piece of a larger puzzle—a puzzle that, when assembled correctly, creates a seamless and intuitive living environment. Embrace the basics, and you'll soon find that the world of smart homes is both accessible and endlessly fascinating.

Chapter 3: The Building Blocks of a Smart Home

In this chapter, we delve into the essential components that form the foundation of any smart home. A solid understanding of these building blocks is crucial, as they ensure that your devices not only function individually but also communicate seamlessly with one another. From connectivity protocols to the nuts and bolts of your home network, this chapter provides a detailed guide to creating a reliable, efficient, and secure smart home environment.

3.1 Connectivity Essentials

Connectivity is the lifeblood of a smart home. It's what allows all your devices to interact, share data, and perform tasks automatically. In this section, we explore the various communication protocols and technologies that make it possible.

3.1.1 Communication Protocols

Smart devices rely on several communication methods, each with its own advantages and trade-offs. Understanding these protocols will help you choose devices that best suit your needs.

Wi-Fi

- **Overview:**
  Wi-Fi is the most common connectivity standard in homes today. It offers high data transfer speeds and

is widely supported across smart devices.

- **Pros:**

    - High bandwidth ideal for data-intensive tasks such as streaming video from security cameras.

    - Ubiquitous and familiar, with almost every home router supporting it.

- **Cons:**

    - Relatively high power consumption compared to some other protocols, which can affect battery-powered devices.

    - Can become congested in homes with many devices connected to a single network.

Bluetooth

- **Overview:**
  Bluetooth is designed for short-range communication and is commonly used for connecting personal devices such as speakers, headphones, and wearable gadgets.

- **Pros:**

    - Low energy consumption, particularly with Bluetooth Low Energy (BLE), making it ideal for devices that run on batteries.

- ○ Simple pairing process.

- **Cons:**

  - ○ Limited range, typically effective within 10–30 meters.

  - ○ Lower data throughput compared to Wi-Fi.

Zigbee and Z-Wave

- **Overview:**
  These are dedicated mesh networking protocols designed for home automation. They enable devices to pass signals through one another, extending the network's effective range.

- **Pros:**

  - ○ Low power consumption, which is perfect for battery-operated devices like sensors and smart locks.

  - ○ Reliable connectivity in a mesh network, as devices can relay signals to one another.

- **Cons:**

  - ○ Lower data rates, which is acceptable for control signals but not for streaming data.

  - ○ Sometimes require a central hub or bridge to integrate with your primary network.

Emerging Protocols: Matter and Thread

- **Matter:**

  - A unified, open-source standard designed to improve device interoperability.

  - Promises to simplify the smart home ecosystem by ensuring that devices from different manufacturers can work together seamlessly.

- **Thread:**

  - A low-power mesh networking protocol built specifically for IoT devices.

  - Offers robust security features and reliable connectivity, making it ideal for smart home applications.

3.1.2 Choosing the Right Protocol for Your Needs

When selecting smart home devices, consider the following:

- **Range:**
  Determine how far the signal needs to travel in your home. Large homes may benefit from mesh networks (Zigbee, Z-Wave, or Thread) to cover every corner.

- **Power Consumption:**
  For battery-operated devices, opt for low-energy

protocols like Bluetooth LE or Zigbee.

- **Data Requirements:**

  For tasks that involve high data transfer (such as streaming security camera footage), Wi-Fi may be the best option.

- **Interoperability:**

  With emerging standards like Matter, look for devices that promise greater compatibility with other brands and ecosystems.

3.2 Home Networking Basics

A robust home network is the backbone of your smart home. This section provides a comprehensive guide on setting up and maintaining a network that can handle multiple devices efficiently and securely.

3.2.1 Setting Up Your Home Network

Selecting the Right Router

- **Key Features:**

  - **Dual-Band or Tri-Band:**
    Routers that support multiple frequency bands (2.4 GHz and 5 GHz) can help reduce interference and manage multiple devices better.

- Coverage Area:
  Consider routers with extended range or mesh systems if you have a large home or experience dead zones.

- Security Features:
  Look for routers that offer robust security options like WPA3 encryption, built-in firewalls, and automatic firmware updates.

Placement of Your Router

- **Optimal Location:**

  - Place your router in a central location to ensure even coverage throughout your home.

  - Avoid obstructions such as thick walls or metal objects that can interfere with the signal.

- **Interference Management:**

  - Keep the router away from other electronic devices that might cause signal interference, such as microwaves or cordless phones.

3.2.2 Enhancing Connectivity

Mesh Networking

- **Overview:**
  Mesh networks use multiple nodes to create a blanket of Wi-Fi coverage, ensuring that every

corner of your home remains connected.

- **Benefits:**

  o Seamless roaming: Devices automatically switch to the strongest node.

  o Scalability: Easily add more nodes to extend coverage as needed.

Wired Connections vs. Wireless

- **Wired Connections:**

  o Offer more stable and faster connections for devices that require high bandwidth, such as smart TVs or gaming consoles.

- **Wireless Connections:**

  o Provide flexibility and ease of installation for devices that are mobile or not near a router.

3.2.3 Securing Your Home Network

Network Security Best Practices

- **Strong Passwords:**
  Use complex passwords for both your Wi-Fi network and router admin panel.

- **Regular Updates:**
  Ensure your router's firmware is updated to protect against vulnerabilities.

- **Guest Networks:**
  Set up a separate network for guests and IoT devices to isolate them from your main network, reducing potential security risks.

- **Encryption:**
  Use WPA3 or WPA2 encryption to safeguard your network data.

Monitoring and Troubleshooting

- **Network Monitoring Tools:**
  Utilize apps and software that help track the performance and security of your network.

- **Troubleshooting Common Issues:**
  Familiarize yourself with basic troubleshooting steps such as rebooting your router, checking for firmware updates, and adjusting channel settings to avoid interference.

3.3 Key Components of a Smart Home System

Beyond connectivity and networking, certain hardware components are essential for bringing your smart home to life. This section explores these critical devices and their roles.

3.3.1 Smart Hubs and Controllers

What is a Smart Hub?

- **Definition:**
  A smart hub acts as the central nervous system of your smart home, connecting and managing various devices and allowing them to communicate with each other.

- **Functions:**

  - **Device Integration:**
    Enables disparate devices to work together by acting as a common interface.

  - **Automation Management:**
    Allows you to create and manage routines or scenes that automate multiple devices simultaneously.

  - **Voice Control Integration:**
    Many hubs integrate with voice assistants (e.g., Amazon Alexa, Google Assistant, Apple Siri) for hands-free control.

Popular Smart Hubs:

- Examples include dedicated devices like Samsung SmartThings Hub or integrated solutions within smart speakers. When choosing a hub, consider compatibility with your existing devices and future expansion plans.

3.3.2 Routers and Bridges

Role of Routers in a Smart Home

- **Connectivity Backbone:**
  Routers distribute internet connectivity to all your smart devices, making them a critical component of your network.

- **Advanced Features:**
  Modern routers may include features like Quality of Service (QoS) to prioritize bandwidth for important devices, ensuring that critical systems like security cameras operate smoothly.

Bridges and Gateways

- **Definition:**
  Bridges or gateways serve as connectors between devices that use different communication protocols. They ensure that devices from varying ecosystems can communicate effectively.

- **Examples:**

  - A Zigbee bridge might be required to connect your Zigbee-enabled sensors to a Wi-Fi network.

  - Some smart home systems include integrated gateways that simplify the setup process.

3.3.3 Additional Components

Smart Switches and Sensors

- **Smart Switches:**

  Replace traditional light switches, allowing you to control lighting remotely or through automation routines.

- **Sensors:**

  Devices such as motion sensors, temperature sensors, and door/window sensors provide real-time data that can trigger automated actions (e.g., turning on lights when someone enters a room).

Smart Plugs and Outlets

- **Functionality:**

  These devices allow you to control power to non-smart appliances, effectively making them part of your smart home network.

- **Energy Monitoring:**

  Many smart plugs also offer energy usage data, helping you monitor and reduce electricity consumption.

Final Thoughts

In this chapter, we've dissected the critical building blocks of a smart home—from the underlying connectivity protocols and home networking basics to the hardware components that make up your system. With a thorough understanding of these elements, you're now equipped to make informed decisions when selecting and setting up your smart home devices. Whether you're focusing on ensuring reliable network coverage or integrating various devices into a cohesive system, these foundational concepts will guide you in creating a smart home that is efficient, secure, and adaptable to future innovations.

As you progress, remember that the strength of your smart home lies in the seamless interplay between these components. In the chapters ahead, we'll build upon this foundation to explore device selection, integration strategies, and advanced automation techniques.

Chapter 4: Choosing Your Smart Home Devices

Selecting the right smart home devices is a critical step in building a system that not only meets your current needs but also has the flexibility to grow with future innovations. In this chapter, we'll provide a comprehensive guide to help you navigate the vast array of options available. We'll cover understanding different smart home ecosystems, evaluating budget-friendly options without compromising quality, and ensuring that your devices are compatible and future-proof.

4.1 Understanding Ecosystems

A smart home ecosystem refers to the collection of devices and platforms that work together under a unified control system. Choosing an ecosystem that fits your lifestyle is key to creating a seamless experience.

Major Platforms and Their Characteristics

- **Amazon Alexa:**

  - **Overview:**
    Amazon's Alexa ecosystem is one of the most popular choices for beginners. It integrates with a wide range of devices—from smart speakers and lights to thermostats and security cameras.

- **Strengths:**
  - Extensive third-party device support.
  - Wide range of skills (apps) that allow for varied functionality.
  - Regular updates and a growing list of compatible products.
- **Considerations:**
  - Some devices might have limited functionality outside the Alexa ecosystem.
  - Privacy concerns for some users due to data collection practices.

- **Google Home:**
  - **Overview:**
  Google Home offers powerful voice search capabilities and seamless integration with Google services. It's ideal for users who rely heavily on Google's ecosystem.
  - **Strengths:**
    - Excellent natural language processing and voice recognition.
    - Deep integration with Google services (Calendar, Maps, etc.).

- Wide compatibility with third-party devices.

  o **Considerations:**

    - As with Alexa, the ecosystem tends to work best when most devices are Google-enabled.

    - Similar privacy considerations as other voice assistants.

- **Apple HomeKit:**

  o **Overview:**

  Apple HomeKit is designed for users within the Apple ecosystem, offering tight integration with iOS devices. It emphasizes privacy and security.

  o **Strengths:**

    - Robust security protocols and privacy features.

    - Intuitive control via the Home app on iOS and Siri integration.

    - Consistent user experience across Apple devices.

- Considerations:
  - A more limited selection of compatible devices compared to Alexa or Google Home.
  - Generally, higher price points and less flexibility for non-Apple products.

Single Ecosystem vs. Mixed-Device Approach

- **Single Ecosystem:**
  - **Benefits:**
    - Simplified setup and management.
    - Consistent user interface and control experience.
    - Optimized automation routines and device integration.
  - **Drawbacks:**
    - May limit device selection to products approved by that ecosystem.
    - Potential vendor lock-in, making future transitions more challenging.

- **Mixed-Device Approach:**
  - **Benefits:**
    - Greater flexibility in choosing devices

based on features and price.

- Ability to select the best device for each function regardless of brand.

- ○ **Drawbacks:**

  - Potential compatibility issues and the need for multiple apps or hubs.

  - Increased complexity in managing different protocols and interfaces.

4.2 Budget-Friendly Options

Building a smart home doesn't have to break the bank. There are plenty of affordable devices that provide significant functionality without the premium price tag.

Essential Devices for Beginners

- **Smart Speakers and Voice Assistants:**

  - ○ Affordable options like the Amazon Echo Dot or Google Nest Mini offer excellent entry points into smart home control.

  - ○ These devices serve as hubs for voice control and can manage multiple smart home devices from a central location.

- **Smart Bulbs:**
  - LED smart bulbs are one of the simplest ways to start automating your home.
  - They typically allow for remote control, scheduling, and color changes, adding both functionality and ambiance.

- **Smart Plugs and Outlets:**
  - These devices enable you to control traditional appliances by turning power on and off remotely.
  - Many smart plugs offer energy monitoring features, helping you track consumption and save on bills.

- **Basic Sensors:**
  - Entry-level sensors (like motion or door/window sensors) are usually affordable and can trigger automated routines for lighting or security.

Evaluating Quality Versus Price

- **Brand Reputation:**
  - Research brands with strong customer reviews and active support communities.
  - Established brands might offer higher upfront

costs but can provide better reliability and longer-term support.

- **Feature Set:**

  o   Identify which features are essential for your needs.

  o   Avoid overpaying for advanced features you might not use immediately; instead, look for products that offer scalable options.

- **Future Expansion:**

  o   Consider whether a budget device can easily integrate with additional devices later.

  o   Sometimes spending a little more upfront can save you from compatibility headaches in the future.

- **Sales and Bundles:**

  o   Keep an eye out for seasonal sales or bundle offers that provide a complete starter kit at a reduced price.

4.3 Compatibility & Future-Proofing

In a rapidly evolving technological landscape, ensuring that your smart home devices remain compatible and can adapt to future advancements is crucial.

Ensuring Compatibility

- **Standard Protocols:**

    o Look for devices that use widely adopted communication protocols like Wi-Fi, Bluetooth, Zigbee, or Z-Wave.

    o The emergence of universal standards like Matter promises to enhance compatibility across brands and devices.

- **Ecosystem Integration:**

    o Verify that devices can integrate with your chosen ecosystem (Alexa, Google Home, HomeKit, etc.).

    o Read product specifications and user reviews to confirm seamless integration and reliable performance.

- **Hub and Bridge Requirements:**

    o Some devices require a specific hub or bridge to operate effectively.

    o Ensure that any additional hardware is compatible with both the device and your existing network setup.

Planning for the Future

- **Scalability:**

  - Choose devices that allow for expansion.

  - A smart home system should be modular—easy to upgrade or add new devices without a complete overhaul.

- **Firmware and Software Updates:**

  - Prioritize brands that have a track record of providing regular firmware and software updates.

  - This practice not only improves performance but also enhances security and compatibility with future devices.

- **Open Standards:**

  - Embrace devices that adhere to open standards.

  - Open standards can reduce vendor lock-in and make it easier to integrate products from different manufacturers as your system evolves.

- **Future Trends:**

  - Keep an eye on emerging technologies like edge computing and AI-driven automation, which may influence the capabilities of smart home devices.

- o When possible, choose devices that are designed to leverage these advancements, ensuring your home remains state-of-the-art.

4.4 Additional Considerations

While understanding ecosystems, budgets, and compatibility is central to choosing your smart home devices, there are a few additional factors that can further enhance your decision-making process.

Aesthetics and Design

- **Home Decor Integration:**
  - o Consider devices that blend well with your home's interior design.
  - o Many manufacturers now offer products in various finishes and styles, ensuring that technology enhances rather than detracts from your living space.
- **User Interface and Control:**
  - o Evaluate the design of companion apps and control interfaces.

- A user-friendly interface can greatly simplify setup and daily operations.

Energy Consumption and Sustainability

- **Efficiency Ratings:**

  - Look for devices with high energy efficiency ratings.

  - Energy-efficient devices can reduce your overall consumption and lower utility bills.

- **Eco-Friendly Options:**

  - Some smart home devices are designed with sustainability in mind, using recyclable materials and low-power operation.

Customer Support and Warranty

- **Service Availability:**

  - Reliable customer support can be invaluable when troubleshooting issues or integrating new devices.

- **Warranty and Return Policies:**

  - Pay attention to warranty periods and return policies.

  - These factors can provide added assurance, especially when investing in new technology.

Final Thoughts

Chapter 4 has provided a detailed roadmap for selecting the smart home devices that best suit your needs, preferences, and budget. By understanding various ecosystems, exploring budget-friendly options, and emphasizing compatibility and future-proofing, you can build a system that is both robust and flexible.

As you choose your devices, remember that the goal is to create a smart home that is not only efficient and secure but also adaptable to the rapidly changing landscape of home automation technology. With thoughtful planning and informed decisions, you can set the stage for a seamless and enjoyable smart home experience.

Chapter 5: Setting Up and Integrating Your Smart Devices

This chapter is dedicated to the practical aspects of creating your smart home. We'll guide you through the process of setting up each device and integrating them into a cohesive system. With detailed, step-by-step instructions, you'll learn how to connect your gadgets to your home network, use companion apps and voice assistants for control, and create custom automations that make your home truly intelligent.

5.1 Step-by-Step Setup Guides

Before you can enjoy the benefits of a smart home, you need to ensure that each device is properly installed and connected. This section outlines a comprehensive, step-by-step approach to help you set up your devices with confidence.

5.1.1 Unpacking and Inspecting Your Devices

- **Unpacking:**
    - Carefully remove each device from its packaging.
    - Check the included manuals, power adapters, mounting hardware, and any quick-start guides.

- **Inspection:**

  - Examine the device for any physical damage that might have occurred during shipping.

  - Verify that all required components are present before beginning installation.

5.1.2 Initial Installation

- **Powering Up:**

  - Connect devices that require a constant power source to an outlet or a smart plug.

  - For battery-powered devices, insert the batteries as instructed and verify they power on (often indicated by LED lights or a sound).

- **Physical Placement:**

  - Follow manufacturer recommendations for optimal placement (e.g., a smart thermostat should be installed away from direct sunlight and drafts, while smart speakers perform best when centrally located).

  - Consider factors such as Wi-Fi signal strength and sensor accuracy when placing devices.

5.1.3 Connecting to Your Home Network

- **Wi-Fi and Network Setup:**

  o Using your smartphone or computer, search for the device's Wi-Fi network (often the device creates a temporary hotspot for initial configuration).

  o Follow the on-screen instructions in the device's companion app to connect the device to your home Wi-Fi network.

  o Enter your Wi-Fi credentials when prompted and wait for the device to confirm a successful connection.

- **Alternative Protocols:**

  o For devices using Bluetooth, Zigbee, or Z-Wave, follow the specific pairing instructions, which may involve placing the device in pairing mode and using a dedicated hub or bridge to complete the setup.

5.1.4 Firmware Updates and Calibration

- **Firmware Checks:**

  o Once connected, check the device settings in the companion app for available firmware or software updates.

  o Update your device if needed to ensure it has

the latest features, security patches, and performance improvements.

- **Calibration and Testing:**

  - Follow any calibration procedures outlined in the manual (for instance, calibrating sensors or adjusting the range of motion detectors).

  - Test the device's basic functions (e.g., turning a smart bulb on/off, verifying temperature settings on a thermostat) to confirm that it is working correctly.

5.2 Using Apps and Voice Assistants

A user-friendly interface is key to managing your smart devices. This section explains how to use the companion apps and voice assistants that come with many smart home systems.

5.2.1 Companion Apps

- **Installation and Account Creation:**

  - Download the manufacturer's app from your device's app store.

  - Create an account if required. Many apps offer a guided tour for first-time users to help set up your device.

- **Navigating the Interface:**

  o Familiarize yourself with the app's dashboard where you can monitor device status, adjust settings, and create automations.

  o Learn to access troubleshooting tips, device logs, and update notifications within the app.

- **Device Grouping and Management:**

  o Organize devices into rooms or groups to simplify control. For example, grouping all kitchen devices together can make it easier to create scene-based automations.

  o Set up notifications and alerts for important events (such as security breaches or low battery warnings).

5.2.2 Voice Assistants

- **Pairing with Your Voice Assistant:**

  o Open your preferred voice assistant's app (Amazon Alexa, Google Assistant, or Apple HomeKit).

  o Follow the in-app instructions to add new devices, which usually involves enabling a specific skill or linking your smart home account.

- **Voice Commands and Routines:**

  o Learn the basic voice commands for your devices. For example, "Alexa, turn off the living room lights" or "Hey Google, set the thermostat to 72°F."

  o Explore voice-controlled routines— predefined sequences that can control multiple devices with a single command (e.g., "Goodnight" might turn off lights, lock doors, and adjust the thermostat).

- **Customization:**

  o Customize your voice assistant settings for preferred names and control preferences to streamline interactions.

  o Regularly update the device lists as you add or remove devices from your system.

5.3 Integration Techniques

Integrating multiple devices into a cohesive system is where the true power of a smart home shines. This section offers strategies and examples to help you link your devices together.

5.3.1 Creating Automations and Routines

- **Understanding Automations:**

  - Automations are sets of instructions that trigger actions based on specific conditions (such as time, sensor input, or user commands).

  - Routines allow multiple automations to run in sequence with a single command.

- **Step-by-Step Automation Creation:**

1. **Identify a Need:**

   - Decide what you want to achieve. For instance, automate your morning routine so that lights gradually brighten, the thermostat adjusts to your preferred temperature, and your coffee maker starts brewing.

2. **Set the Trigger:**

   - Choose a trigger for your automation, such as a specific time, motion detection, or a voice command.

3. **Define the Actions:**

   - Select the devices that should respond and set the actions they should take. Use the companion app to add each action sequentially.

4. **Test and Adjust:**

    - Run the automation to ensure it works as expected, then adjust timings or actions if necessary.

- **Examples of Common Automations:**

    o **Morning Routine:** Gradually brighten lights, adjust thermostat, and start music as you wake up.

    o **Away Mode:** Automatically lock doors, adjust lighting, and arm security systems when you leave the house.

    o **Energy Saver:** Turn off non-essential devices when no movement is detected in a room for a specified period.

5.3.2 Synchronizing Devices Across Platforms

- **Multi-Platform Integration:**

    o In mixed-device environments, use bridging devices or platforms (like IFTTT, SmartThings, or Home Assistant) to create interoperability between devices from different ecosystems.

    o These platforms can help create custom integrations where native compatibility is lacking.

- **Using IFTTT (If This Then That):**

  - **Creating Applets:**

    - Learn to create "applets" that connect actions between different devices. For example, if your smart doorbell detects motion, it could trigger your smart lights to turn on.

  - **Customization:**

    - Explore pre-built applets available in the IFTTT community and modify them to suit your needs.

- **Platform-Specific Tools:**

  - Many ecosystems offer built-in integration tools that help you connect devices without needing third-party services. Explore the automation sections of your preferred smart home app to discover available options.

5.3.3 Troubleshooting Integration Issues

- **Common Challenges:**

  - Devices may fail to connect if they're on different networks or if firmware is outdated.

  - Incompatibility between devices from

different manufacturers can occasionally cause communication issues.

- **Diagnostic Steps:**

1. **Verify Connectivity:**

    - Ensure all devices are connected to the same network or properly linked via hubs/bridges.

2. **Check Firmware and Updates:**

    - Make sure every device and its companion app are running the latest software versions.

3. **Reboot and Re-pair:**

    - A simple restart or re-pairing of devices can resolve many integration issues.

4. **Consult Support Resources:**

    - Use the troubleshooting sections within apps or visit manufacturer websites for additional guidance.

- **Building an Integration Toolkit:**

    o Maintain a checklist of common troubleshooting steps.

    o Keep a record of each device's configuration

details to quickly re-establish connections if needed.

Final Thoughts

In Chapter 5, we've covered the full lifecycle of setting up and integrating your smart devices—from the unboxing and initial configuration to harnessing the power of companion apps, voice assistants, and custom automations. With these detailed instructions, you should be well-equipped to transform individual devices into a fully interconnected smart home system.

By following these guidelines, you not only ensure that your devices work efficiently on their own but also in harmony with each other. This integration is what creates a seamless, user-friendly experience, allowing you to enjoy the convenience and enhanced control that a smart home offers.

Chapter 6: Automating Your Home

Automation is the heartbeat of a smart home. It transforms isolated devices into a synchronized ecosystem that responds to your needs, anticipates your routines, and creates a seamless living experience. In this chapter, we'll explore everything from the fundamentals of automation to advanced strategies that empower you to tailor your smart home exactly how you want it. Whether you're a beginner looking to automate basic tasks or an enthusiast eager to delve into multi-step routines, this chapter provides detailed, step-by-step guidance to help you get the most out of your smart home.

6.1 Understanding Automation

Automation in the smart home context involves programming devices to perform actions based on triggers, conditions, or schedules. By automating tasks, you reduce manual intervention, improve energy efficiency, and enhance the overall convenience of your living space.

What Is Automation?

- **Definition:**
  Automation is the process of setting predefined actions (or sequences of actions) to occur automatically when certain conditions or triggers are met. This can involve a single device or multiple

devices acting in concert.

- **Key Components:**

  - **Triggers:**
    Events or conditions that set the automation in motion. These can be time-based (e.g., at 7:00 AM), sensor-based (e.g., when motion is detected), or location-based (e.g., when you leave the house).

  - **Actions:**
    The tasks that are executed when the trigger occurs. Actions can include turning lights on/off, adjusting the thermostat, or sending notifications.

  - **Conditions:**
    Optional parameters that refine when an automation should run. For example, an automation might only execute if it's after sunset or if the ambient temperature exceeds a certain threshold.

Types of Automation

- **Scenes:**
  Predefined settings that adjust multiple devices simultaneously. For example, a "Movie Night" scene might dim the lights, close the curtains, and turn on the TV.

- **Schedules:**
Time-based routines that run automatically at set times. This could include a morning routine that gradually brightens lights and starts your coffee maker.

- **Routines:**
Multi-step processes that can combine several triggers, actions, and conditions. Routines can be customized for various activities such as leaving home, arriving, or even for specific events like a party.

Benefits of Automation

- **Convenience:**
Automations reduce the need for manual intervention, allowing you to focus on other tasks.

- **Efficiency:**
Automated actions optimize energy use—for example, adjusting lighting and temperature only when necessary.

- **Security:**
Automations can enhance home security by triggering alarms or notifications in response to unusual activity.

- **Personalization:**
By tailoring automation routines to your lifestyle, your home becomes more responsive and aligned

with your daily habits.

6.2 Creating Your Own Automations

This section walks you through designing and implementing your own automation routines, offering a detailed guide that covers planning, execution, and refinement.

Step 1: Identify the Task or Need

- **Brainstorm Tasks:**
  Think about daily activities or problems that could be simplified with automation. For example, you might want the lights to turn on automatically when you enter a room.

- **Prioritize:**
  Start with simple tasks to build confidence before tackling more complex routines.

Step 2: Define Triggers, Actions, and Conditions

- **Select a Trigger:**
  Choose a condition that will initiate your automation. Common triggers include:

  - **Time-Based:** Specific times of the day (e.g., "at 7:00 AM").

  - **Sensor-Based:** Input from motion sensors, door sensors, or temperature sensors.

- o **Location-Based:** Your phone's GPS can trigger actions when you leave or arrive at home.

- **Decide on Actions:**
  Identify the devices or systems that will respond. For example:

  - o Turning on/off lights.

  - o Adjusting the thermostat.

  - o Activating smart plugs or appliances.

- **Set Conditions (if needed):**
  Refine your automation by adding conditions. For instance, you may want the automation to work only if the home is dark (using a light sensor) or only on weekdays.

Step 3: Use Companion Apps or Voice Assistants to Program the Automation

- **Companion Apps:**
  Most smart home devices come with a companion app that includes an automation or routines section. These apps offer user-friendly interfaces to help you build and test automations.

- **Voice Assistants:**
  Platforms like Amazon Alexa, Google Assistant, or Apple HomeKit allow you to set up automations using voice commands or through their respective

apps.

- **Walkthrough Example:**
  Imagine you want to create a "Good Morning" routine:

  1. **Trigger:** Set the trigger for 7:00 AM.

  2. **Actions:**

     - Gradually brighten the bedroom lights.

     - Adjust the thermostat to a comfortable morning temperature.

     - Start playing your favorite morning playlist on a smart speaker.

  3. **Conditions:**

     - Only execute if it's a weekday.

  4. **Testing:**
     Run the routine manually from the app to confirm that all devices respond correctly, then enable the automation.

Step 4: Test and Refine

- **Trial Runs:**
  Execute the automation manually to ensure each component works as intended.

- **Monitor and Adjust:**
  Observe the automation's performance over several

days and make adjustments to timings, triggers, or actions as needed.

- **Document Settings:**
  Keep a record of your automations and settings. This helps in troubleshooting and scaling your routines over time.

6.3 Practical Examples of Home Automation

To illustrate the concepts in action, here are several detailed examples of automation routines that you can customize to suit your needs.

Example 1: The Morning Routine

- **Objective:**
  Ease into your day with minimal manual intervention.

- **Trigger:**
  Scheduled to start at 7:00 AM on weekdays.

- **Actions:**

  - **Lighting:** Gradually increase brightness in the bedroom and kitchen to simulate sunrise.

  - **Thermostat:** Adjust to a comfortable temperature before you get out of bed.

  - **Smart Speaker:** Start playing a pre-selected

playlist or a morning news briefing.

- o **Coffee Maker:** If connected, begin brewing coffee at the scheduled time.

- **Conditions:**
Only activate on weekdays and when the ambient light is below a certain level (if your system includes a light sensor).

Example 2: Security Routine When Leaving Home

- **Objective:**
Ensure that your home is secure when you're not around.

- **Trigger:**
Activated when your phone's location indicates that you're leaving the vicinity of your home.

- **Actions:**

  - o **Locks:** Automatically lock all smart locks.

  - o **Security Cameras:** Activate recording or send a status alert.

  - o **Lighting:** Turn off interior lights and turn on exterior security lights.

  - o **Notifications:** Send an alert to your smartphone confirming that the security routine is active.

- **Conditions:**
  Execute only if the home is not already in "Away Mode" and during nighttime hours for additional security.

Example 3: Energy Saver Routine

- **Objective:**
  Reduce energy consumption when rooms are not in use.

- **Trigger:**
  Activated by motion sensors that detect no movement in a room for a set period (e.g., 10 minutes).

- **Actions:**

  - **Lights:** Automatically turn off the lights.

  - **Smart Plugs:** Power down devices that are not critical.

  - **Thermostat:** Adjust HVAC settings to an energy-saving mode.

- **Conditions:**
  Ensure the routine doesn't activate during certain hours (e.g., when you're likely to be home and active).

6.4 Using Automation Tools and Platforms

While many devices offer built-in automation capabilities through their companion apps, third-party platforms can further enhance your control and integration options.

Popular Automation Platforms

- **IFTTT (If This Then That):**

    o **Functionality:**
      IFTTT allows you to create applets that connect devices and services from different ecosystems.

    o **Usage:**

        ▪ For instance, set up an applet where if your smart doorbell detects motion, your smart lights automatically turn on.

    o **Advantages:**

        ▪ User-friendly interface and extensive community support.

- **SmartThings:**

    o **Functionality:**
      Samsung's SmartThings hub provides a centralized platform for managing a wide range of smart devices.

- ○ **Usage:**

  - ▪ Create sophisticated routines that integrate devices from various brands.

- ○ **Advantages:**

  - ▪ Robust integration features and compatibility with many protocols.

- **Home Assistant:**

  - ○ **Functionality:**
    An open-source platform that offers deep customization and integration options.

  - ○ **Usage:**

    - ▪ Ideal for users who want to build complex automations and have granular control over every aspect of their smart home.

  - ○ **Advantages:**

    - ▪ Highly flexible, though it requires more technical know-how.

Choosing the Right Platform

- **Compatibility:**
  Ensure the platform supports the devices you use.

- **Ease of Use:**
  Beginners might prefer the simplicity of IFTTT or

native apps, while advanced users might opt for Home Assistant.

- **Scalability:**
  Consider how easily the platform can grow with your expanding smart home system.

6.5 Advanced Automation Strategies

For those looking to push the boundaries of what their smart home can do, this section explores advanced strategies and techniques.

Multi-Step Routines with Conditional Logic

- **Complex Automations:**
  Combine multiple triggers, conditions, and actions into one routine. For example, you can create an automation that:

  - Turns on outdoor lights when motion is detected,

  - Checks the time of day,

  - And only activates if you haven't manually overridden the setting.

- **Using Conditional Logic:**
  Some platforms allow "if-then-else" logic, where the system can perform different actions based on varying conditions. This is especially useful for

security or energy-saving scenarios.

Voice Assistant Integration

- **Enhanced Control:**
  Use voice commands to initiate or modify routines on the fly. For example, saying "Goodnight" might trigger a multi-step routine that locks doors, dims lights, and sets the thermostat to an energy-saving temperature.

- **Custom Commands:**
  Many voice assistants let you define custom routines and aliases, providing a more natural interaction with your home.

Troubleshooting and Refining Advanced Automations

- **Common Issues:**

  - Automations might not trigger due to network issues, outdated firmware, or conflicts between devices.

  - Debugging requires verifying that each device in the routine is responsive.

- **Best Practices:**

  - Regularly update firmware and review automation logs available in many companion apps.

  - Use a systematic approach: test each

component individually before integrating them into a larger routine.

- ○ Document your routines and settings, so you can revert or tweak them as needed.

Final Thoughts

Automating your home is about creating a living space that adapts to your lifestyle with minimal effort on your part. By understanding the core principles of automation, following a structured approach to creating your routines, and leveraging powerful automation platforms, you can transform your home into an intelligent, responsive environment.

As you experiment with automations—whether simple or complex—you'll discover new ways to enhance convenience, improve security, and save energy. Remember that every successful automation starts with a clear objective, thorough testing, and ongoing refinement. Embrace the process, and let your smart home evolve to meet your unique needs.

Chapter 7: Troubleshooting Common Issues

Even the most well-planned smart home can encounter occasional hiccups. In this chapter, we address the common issues you may face, provide practical troubleshooting tips, and guide you in building a toolkit to diagnose and resolve problems. Whether you're dealing with connectivity lapses, device pairing failures, or firmware glitches, this detailed guide will empower you to keep your smart home running smoothly.

7.1 Identifying Common Problems

Before you can fix an issue, it's important to accurately identify it. This section describes the most frequently encountered problems and helps you understand their symptoms.

7.1.1 Connectivity Issues

- **Symptoms:**
    - Devices failing to respond to commands.
    - Intermittent connections or "dead zones" in your home.
    - Slow response times or timeouts when accessing device controls via an app.

- **Potential Causes:**

  - Weak Wi-Fi signals or interference from other electronic devices.

  - Overloaded networks, especially during peak usage times.

  - Incorrect network configurations or outdated router firmware.

- **Diagnostic Questions:**

  - Are all devices experiencing issues or just specific ones?

  - Have you recently changed your network settings or added new devices?

  - Is the problem persistent or does it vary by time or location?

7.1.2 Device Pairing Failures

- **Symptoms:**

  - Devices not appearing in companion apps.

  - Repeated prompts to pair or connect.

  - Failure to recognize a device after initial setup.

- **Potential Causes:**

  - Interference from nearby devices.

  - Outdated firmware on either the device or the hub/bridge.

  - Incompatible protocols or issues with the device's pairing mode.

- **Diagnostic Questions:**

  - Is the device properly powered and in pairing mode?

  - Have you checked for firmware or app updates?

  - Are you following the manufacturer's recommended pairing procedure?

7.1.3 Firmware Glitches and Software Bugs

- **Symptoms:**

  - Devices suddenly becoming unresponsive or restarting unexpectedly.

  - Features malfunctioning even when the device appears to be connected.

  - Error messages or unusual behavior after an update.

- **Potential Causes:**

  - Outdated or corrupted firmware.

  - Incompatibility issues following a recent software update.

  - Conflicts between third-party apps and device software.

- **Diagnostic Questions:**

  - When did the issue first appear? Was it after a firmware update?

  - Is the problem isolated to one device or affecting multiple devices from the same manufacturer?

  - Have you tried resetting the device to its default settings?

7.2 Practical Troubleshooting Tips

With a clear understanding of common issues, you can now apply these step-by-step troubleshooting tips to resolve problems as they arise.

7.2.1 Step-by-Step Troubleshooting Process

1. **Restart Devices and Network Equipment:**

   - **Simple Reset:** Begin by turning off and unplugging the affected device, waiting a minute, and powering it back on.

- o **Router Reset:** Restart your router and modem to clear temporary network issues.

- o **Power Cycle Hubs/Bridges:** If you're using a hub or bridge, reboot it to re-establish connections.

2. **Check Device Settings and Configurations:**

- o **Network Settings:** Verify that the device is connected to the correct Wi-Fi network and that signal strength is adequate.

- o **Pairing Mode:** Ensure the device is in pairing mode, as indicated by its LED status or instructions in the manual.

- o **User Credentials:** Confirm that you're using the correct login details for the device's companion app.

3. **Update Firmware and Software:**

- o **Device Firmware:** Use the companion app to check for firmware updates and install any available patches.

- o **Router/Hub Firmware:** Update your router or hub to the latest version to benefit from enhanced stability and security.

- o **App Updates:** Make sure the companion apps for your devices are up-to-date, as developers frequently address bugs and compatibility issues.

4. **Review Automation and Integration Settings:**

   - o **Automation Conflicts:** Disable recently created automations to see if they're interfering with device performance.

   - o **Multi-Device Interference:** Ensure that devices using similar frequencies (e.g., Bluetooth, Zigbee) are not causing cross-talk or interference.

5. **Reset to Factory Settings (if necessary):**

   - o **Last Resort:** If a device continues to malfunction despite troubleshooting, consider resetting it to factory settings.

   - o **Backup Configurations:** Before doing this, back up any custom configurations or automations, as a reset will often clear these settings.

6. **Consult Documentation and Support Resources:**

   - o **User Manuals:** Refer to the device's manual or online support pages for specific troubleshooting guides.

- ○ **Community Forums:** Many manufacturers maintain user forums where you can find advice from other smart home enthusiasts.

- ○ **Customer Support:** If all else fails, contact the manufacturer's support team for guidance or repair options.

7.2.2 Common Troubleshooting Scenarios

- **Scenario 1: A Smart Bulb Doesn't Respond:**

  - ○ Verify the bulb is powered and properly screwed into its socket.

  - ○ Ensure it's connected to the same network as your controller.

  - ○ Check the bulb's status in the companion app and attempt a firmware update.

  - ○ If unresponsive, reset the bulb (following the manufacturer's reset procedure) and re-pair it.

- **Scenario 2: A Smart Thermostat Loses Connection:**

  - ○ Confirm that the thermostat is within range of your Wi-Fi signal and not obstructed by interference.

  - ○ Restart the thermostat and your router.

- Check for software updates and verify that any automations aren't causing conflicts.

- If the issue persists, consider a factory reset or consult customer support.

- **Scenario 3: Voice Assistant Fails to Recognize a Command:**

  - Ensure that the voice assistant's settings are properly configured and that it recognizes your voice.

  - Check the connectivity between the voice assistant and the smart devices.

  - Update both the voice assistant's firmware and the companion app for the device in question.

  - Test commands with different phrasing or reconfigure custom routines if necessary.

7.3 Building a Troubleshooting Toolkit

A proactive approach to troubleshooting involves having the right tools and resources at your fingertips. Here are essential items to include in your smart home troubleshooting toolkit:

7.3.1 Digital Tools and Apps

- **Network Analyzers:**
  Tools like Wi-Fi analyzers can help you assess signal strength, interference, and channel usage throughout your home.

- **Companion App Diagnostics:**
  Many smart home apps include built-in diagnostics, logs, or error messages that can pinpoint issues. Familiarize yourself with these features.

- **Firmware and Software Update Trackers:**
  Use apps or calendar reminders to check for firmware updates regularly. Staying current reduces the risk of encountering known bugs.

7.3.2 Documentation and Checklists

- **Device Manuals and Online Guides:**
  Keep digital or printed copies of user manuals and quick-start guides for all your devices.

- **Troubleshooting Checklists:**
  Create a checklist that includes common troubleshooting steps (restart, update, check connections) for quick reference during an issue.

- **Configuration Records:**
  Document network settings, device configurations, and automation routines. This record can be invaluable if you need to reconfigure devices after a

reset.

7.3.3 Community and Professional Resources

- **Online Forums and User Groups:**
  Participate in communities such as manufacturer forums, Reddit groups, or smart home enthusiast sites. These communities can offer insights and workarounds for uncommon issues.

- **Customer Support Contacts:**
  Maintain a list of customer support contacts for each device manufacturer. Knowing when to escalate an issue can save time and frustration.

- **Professional Help:**
  For persistent or complex issues, consider consulting a professional smart home installer or technician who can provide hands-on support.

Final Thoughts

Troubleshooting is an integral part of maintaining a robust and efficient smart home. Chapter 7 has equipped you with the knowledge to identify common issues, follow practical troubleshooting steps, and build a toolkit that simplifies the process of diagnosing and resolving problems.

Remember that a systematic approach—starting with basic resets and updates, followed by targeted diagnostics—can resolve most issues without the need for professional intervention. As you gain experience, you'll become more adept at pinpointing and solving problems quickly, ensuring that your smart home continues to operate seamlessly.

Chapter 8: Securing Your Smart Home

As smart homes become increasingly integrated into our daily lives, security and privacy become paramount concerns. In this chapter, we focus on safeguarding your smart home by understanding potential risks, implementing best practices, and applying practical security measures. This comprehensive guide will help you build a secure environment that not only protects your data but also ensures your devices function safely and reliably.

8.1 Understanding Security Risks

Before implementing security measures, it's important to understand the potential risks associated with smart home technology.

8.1.1 Common Vulnerabilities

- **Default Credentials:**
  Many smart devices come with default usernames and passwords, which can be easily exploited if not changed promptly.

- **Unsecured Networks:**
  A weak Wi-Fi network can provide an entry point for hackers. Devices on the same network are vulnerable if the network is compromised.

- **Outdated Firmware:**
  Devices that are not regularly updated may contain known vulnerabilities that malicious actors can exploit.

- **Interoperability Issues:**
  Integrating devices from multiple manufacturers can sometimes lead to unforeseen security gaps due to varying security standards.

- **Data Privacy Concerns:**
  Many devices collect data on user behavior, which, if not properly secured, may be accessed by unauthorized parties.

8.1.2 The Impact of Security Breaches

- **Personal Data Exposure:**
  Unauthorized access could lead to the theft of sensitive information such as your daily routines, location data, or even video feeds from security cameras.

- **Control of Devices:**
  A breach might allow hackers to control smart locks, thermostats, lights, or other critical devices, potentially endangering your home's security.

- **Network Compromise:**
  Once a device on your network is compromised, attackers may use it as a foothold to access other devices, leading to a broader network security

issue.

- **Financial and Legal Ramifications:**
  Security breaches can result in financial loss, damage to reputation, and in some cases, legal consequences, especially if personal data is misused.

8.2 Best Practices for Smart Home Security

Implementing robust security measures is essential to protect your smart home. Here are some best practices to help you minimize risks and secure your environment.

8.2.1 Change Default Credentials

- **Immediate Update:**
  Upon installation, change the default usernames and passwords for all devices. Use strong, unique passwords for each device.

- **Password Management:**
  Consider using a reputable password manager to generate and store complex passwords securely.

8.2.2 Secure Your Home Network

- **Strong Encryption:**
  Ensure your Wi-Fi network uses WPA3 (or at least WPA2) encryption to protect data transmissions.

- **Network Segmentation:**
  Create separate networks or guest networks for your smart devices. Isolating IoT devices from your primary network can reduce the risk of a breach spreading.

- **Regular Updates:**
  Routinely check and update your router's firmware and other network equipment to benefit from the latest security patches.

8.2.3 Firmware and Software Updates

- **Automatic Updates:**
  Enable automatic updates where possible to ensure your devices always have the latest security fixes.

- **Manual Checks:**
  Regularly check the manufacturer's website or app for firmware updates if automatic updates aren't available.

- **Vendor Reputation:**
  Purchase devices from reputable brands known for providing timely updates and robust security measures.

8.2.4 Monitor Device Activity

- **Logging and Alerts:**
  Use companion apps or network monitoring tools to keep track of unusual activity or unauthorized

access attempts.

- **Routine Audits:**
  Periodically review connected devices and remove any that are no longer in use or have outdated security.

8.3 Practical Security Measures

Beyond best practices, specific practical measures can further enhance the security of your smart home.

8.3.1 Setting Up Secure Configurations

- **Disable Unnecessary Features:**
  Turn off features or services you don't use (such as remote access) to reduce potential attack surfaces.

- **Two-Factor Authentication (2FA):**
  Enable 2FA on accounts associated with your smart home devices to add an extra layer of security.

- **Custom Network Settings:**
  Change the default SSID (network name) of your Wi-Fi and disable network discovery features to make it harder for intruders to target your network.

8.3.2 Using Guest Networks and Firewalls

- **Guest Network:**
  Create a separate guest network specifically for smart devices. This limits the potential damage if

one device is compromised.

- **Firewalls:**
Implement a firewall on your router or through dedicated security appliances to monitor and control incoming and outgoing network traffic.

- **VPNs for Remote Access:**
If you require remote access to your home network, use a Virtual Private Network (VPN) to secure the connection.

8.3.3 Physical Security Considerations

- **Device Placement:**
Position security cameras and sensors in locations that minimize tampering risks. Ensure that critical devices are not easily accessible to outsiders.

- **Secure Mounting:**
Physically secure devices with mounts or locking mechanisms where applicable, especially for devices in public or accessible areas.

- **Regular Maintenance:**
Regularly inspect devices for signs of physical tampering or damage, and address any issues immediately.

Final Thoughts

Securing your smart home is a continuous process that requires vigilance and proactive management. By understanding the security risks, following best practices, and applying practical measures, you can significantly reduce vulnerabilities and protect your home from potential threats. As you add new devices and expand your system, always consider security first—ensuring that convenience does not come at the cost of safety.

With these detailed guidelines, you're well-equipped to create a secure and resilient smart home environment. Remember that ongoing education and routine security audits are key to keeping your smart home protected as technology evolves.

Chapter 9: Expanding and Future-Proofing Your Smart Home

As technology continues to evolve at a rapid pace, your smart home should not be a static installation but a dynamic system that adapts to new advancements. In this chapter, we explore strategies for expanding your smart home, integrating emerging technologies, and ensuring that your system remains compatible and resilient over time. This comprehensive guide will help you plan for scalability, incorporate future innovations, and make practical decisions that keep your smart home state-of-the-art.

9.1 Planning for Expansion

Before you add new devices or functionalities, it's essential to have a clear plan for how your smart home system will evolve. A scalable design not only maximizes your current investments but also minimizes disruptions when new technology becomes available.

9.1.1 Assessing Your Current System

- **Inventory and Documentation:**
  Start by creating a detailed inventory of all your current devices, including their specifications, communication protocols, and integration points. Document network settings, automation routines, and any custom configurations you've

implemented. This record will serve as a baseline for future additions.

- **Identifying Bottlenecks:**
Evaluate the performance of your existing network and control hubs. Are there areas with weak Wi-Fi signals or devices that frequently experience connectivity issues? Recognizing these bottlenecks early on will help you address them before they affect the integration of new technology.

9.1.2 Planning for Future Additions

- **Modular Approach:**
Adopt a modular strategy where devices and systems can be added or upgraded without requiring a complete overhaul. For instance, use smart hubs and bridges that support multiple protocols, allowing you to integrate devices from various manufacturers.

- **Flexible Infrastructure:**
Consider investing in a robust home network that can accommodate additional devices. Mesh networks and routers with advanced Quality of Service (QoS) settings can help manage increased traffic and ensure reliable connectivity as your smart home grows.

- **Scalability Roadmap:**
Develop a roadmap that outlines your expansion

goals. Identify which rooms or systems (security, lighting, energy management, etc.) you plan to upgrade, and set milestones for when and how these upgrades will occur. This roadmap can be revisited and adjusted as new technology emerges.

9.2 Integrating Emerging Technologies

New technological trends—such as artificial intelligence, edge computing, and next-generation communication protocols—are continuously shaping the smart home landscape. Understanding these trends can help you make informed decisions about future integrations.

9.2.1 Embracing AI and Machine Learning

- **Enhanced Automation:**
  AI-powered devices can learn from your behavior patterns, enabling more intelligent automation routines. For example, smart thermostats that adapt to your schedule or lighting systems that adjust based on activity can improve both convenience and energy efficiency.

- **Predictive Maintenance:**
  With machine learning, your system can predict and alert you to potential issues before they become critical. This proactive approach can reduce downtime and extend the lifespan of your devices.

9.2.2 The Role of Edge Computing

- **Local Data Processing:**
  Edge computing shifts data processing from the cloud to local devices. This change can significantly reduce latency, improve response times, and enhance data privacy by keeping sensitive information on-site.

- **Improved Resilience:**
  As more devices process data locally, your system becomes less dependent on external networks. This decentralization can be particularly useful in maintaining functionality during internet outages or network congestion.

9.2.3 Adopting Next-Generation Protocols

- **Matter and Interoperability:**
  The emergence of universal standards like Matter aims to simplify device compatibility across different ecosystems. As Matter gains traction, consider integrating devices that support it to ensure seamless communication and interoperability.

- **Future-Ready Hubs:**
  Look for smart hubs and bridges that are designed to accommodate firmware updates and support new protocols. These hubs provide a flexible foundation for incorporating upcoming devices without a

complete system overhaul.

9.3 Practical Strategies for Future-Proofing

Ensuring that your smart home remains effective over the long term requires proactive measures. Here are some practical strategies to keep your system current and resilient.

9.3.1 Regular Maintenance and Updates

- **Firmware and Software Updates:**
  Consistently check for and install firmware updates for all your devices, routers, and hubs. Manufacturers often release updates to improve performance, patch security vulnerabilities, and add new features.

- **Periodic System Audits:**
  Conduct regular audits of your smart home system. Review your network performance, update documentation, and verify that all devices are operating optimally. These audits can help you catch issues early and make necessary adjustments before problems escalate.

9.3.2 Investing in Interoperable Devices

- **Open Standards and Compatibility:**
  Prioritize devices that adhere to open standards and are known for their compatibility with multiple

ecosystems. This approach minimizes vendor lock-in and offers greater flexibility when integrating new technologies.

- **Scalable Platforms:**
Choose platforms and hubs that are built with scalability in mind. Research products with a strong track record of supporting updates and integrating a wide range of devices.

9.3.3 Keeping an Eye on Emerging Trends

- **Stay Informed:**
Follow industry news, join smart home communities, and participate in online forums to keep abreast of the latest trends and technological breakthroughs. This knowledge will help you anticipate future needs and make proactive decisions about system expansion.

- **Vendor Roadmaps:**
Review the future roadmaps provided by device manufacturers and service providers. Understanding their planned developments can help you time your purchases and integrations more strategically.

- **Experimentation and Prototyping:**
Don't be afraid to experiment with new devices or technologies on a small scale before integrating them into your primary system. Setting up a test

environment can help you assess the benefits and challenges of new integrations without risking the stability of your core system.

Final Thoughts

Expanding and future-proofing your smart home is an ongoing journey that involves planning, adaptability, and a willingness to embrace new technologies. By designing a modular system, staying informed about emerging trends, and investing in interoperable, scalable devices, you can ensure that your smart home remains efficient, secure, and ahead of the curve.

This chapter has provided you with the strategies and practical advice needed to evolve your smart home system over time. As you implement these strategies, you'll be well-prepared to integrate new technologies seamlessly and maintain a cutting-edge smart home environment that grows with your needs.

Chapter 10: Hands-On Projects and DIY Integrations

This chapter puts theory into practice. It's designed for those who want to roll up their sleeves and dive into real-world projects that bring smart home technology to life. Whether you're an absolute beginner or a seasoned enthusiast, these detailed, step-by-step projects will help you gain confidence and foster creativity while building a more personalized smart home. From automating lighting systems to managing energy consumption and enhancing home security, each project is presented with clear instructions, necessary tools, and troubleshooting tips to ensure success.

10.1 Overview of DIY Projects

Before you begin, it's important to plan your projects and gather the necessary tools and materials. This section provides an overview of the projects featured in this chapter and outlines some general preparatory steps.

10.1.1 Project Preparation

- **Assess Your Needs:**
  Consider what aspects of your home could benefit most from smart automation. Do you want to save energy, improve security, or simply add convenience to daily routines?

- **Gather Tools and Materials:**
  Create a checklist of items you'll need, such as smart bulbs, plugs, sensors, a microcontroller (if applicable), wiring tools, and a smartphone or computer to access companion apps.

- **Set Up a Test Environment:**
  If possible, create a small, controlled environment (like a single room or a designated workspace) to experiment and test your projects before full-scale implementation.

- **Review Safety Guidelines:**
  Ensure you understand basic electrical safety and the proper handling of tools and devices. Always consult the manufacturer's instructions for any equipment used.

10.1.2 Project Management Tips

- **Documentation:**
  Keep detailed notes on your setup, including device configurations, wiring diagrams, and any custom scripts you write. This documentation will be invaluable for troubleshooting and future expansions.

- **Incremental Building:**
  Start with simpler projects and gradually work up to more complex integrations. This step-by-step approach helps build confidence and ensures that

each component works properly before integrating it into a larger system.

- **Community Resources:**
  Engage with online forums, tutorials, and DIY smart home communities. Sharing your progress and learning from others can provide fresh insights and solutions to common challenges.

10.2 Project 1: Smart Lighting Control

This project focuses on automating your home's lighting system, providing enhanced convenience, energy efficiency, and ambiance control.

10.2.1 Objectives

- Automate lighting to respond to time-of-day, motion, or ambient light conditions.

- Enable remote and voice-controlled lighting adjustments.

- Create scenes (e.g., "Movie Mode" or "Dinner Party") that adjust brightness and color for specific moods.

10.2.2 Materials and Tools

- Smart LED bulbs or smart light strips.

- A compatible smart hub or a Wi-Fi-enabled

controller.

- Motion or ambient light sensors (optional for added automation).

- Smartphone with the companion app for your smart lighting system.

- Basic tools (screwdriver, smartphone, and access to your Wi-Fi network).

10.2.3 Step-by-Step Guide

1. **Installation:**

   o Replace traditional bulbs with smart LED bulbs. If using light strips, mount them according to the manufacturer's instructions.

   o Ensure that the devices are powered and within range of your Wi-Fi network.

2. **Configuration:**

   o Download and install the companion app for your smart lighting system.

   o Follow the app's instructions to add each smart bulb or light strip to your network.

   o If using sensors, connect them to your hub and configure their placement (e.g., near entryways or in living areas).

3. **Setting Up Automations:**

- Create a "Sunrise" routine where lights gradually brighten at a set time in the morning.

- Configure a "Motion-Activated" routine that turns lights on when movement is detected and off after a period of inactivity.

- Experiment with color settings to create mood-based scenes (e.g., warm hues for relaxation or cool whites for productivity).

4. **Testing and Troubleshooting:**

- Test each automation individually. For instance, trigger the motion sensor manually and verify that the lights respond as expected.

- Use the app's diagnostic features to monitor connectivity and performance.

- Adjust sensor sensitivity and automation timing based on your observations.

10.2.4 Enhancements

- **Voice Control Integration:**
  Pair your smart lighting system with a voice assistant (Amazon Alexa, Google Assistant, or Apple HomeKit) for hands-free control.

- **Advanced Scheduling:**
  Incorporate seasonal changes or varying routines for weekdays versus weekends.

10.3 Project 2: Energy Management

This project is designed to help you monitor and manage energy usage within your home, contributing to both cost savings and environmental sustainability.

10.3.1 Objectives

- Track energy consumption in real time.

- Automate the powering down of non-essential devices during periods of inactivity.

- Use data to optimize energy use and reduce utility bills.

10.3.2 Materials and Tools

- Smart plugs or smart outlets with energy monitoring capabilities.

- A central hub or a dedicated smartphone/tablet for data visualization.

- Optionally, smart sensors (temperature, motion) to automate energy-saving routines.

- Access to energy monitoring software or companion apps provided by the device manufacturer.

10.3.3 Step-by-Step Guide

1. **Installation:**

   o Replace standard plugs with smart plugs in areas where you want to monitor energy consumption (e.g., living room electronics, kitchen appliances).

   o Ensure the plugs are connected and integrated with your Wi-Fi network.

2. **Configuration:**

   o Use the companion app to set up each smart plug and enable energy monitoring features.

   o Configure any associated sensors to detect occupancy or temperature changes that can trigger energy-saving actions.

3. **Data Collection and Analysis:**

   o Access real-time energy consumption data through the app.

   o Set up alerts for unusual energy usage patterns (e.g., if an appliance is consuming power when it shouldn't be).

4. **Automation Routines:**

   o Create an "Energy Saver" routine that turns off non-critical devices when no movement is detected in a room for a predetermined time.

- o Program smart plugs to activate during off-peak hours, if supported by your utility provider.

5. **Testing and Optimization:**

- o Monitor the energy data over several days and adjust the automation timings and sensor thresholds based on the collected information.

- o Experiment with different settings to maximize energy efficiency without sacrificing comfort.

10.3.4 Enhancements

- **Integrate with Utility Programs:**
Some energy providers offer incentives or dynamic pricing. Configure your system to take advantage of these programs by reducing consumption during peak times.

- **Expand Monitoring:**
Consider integrating additional sensors or smart meters to cover more areas of your home for a comprehensive energy audit.

10.4 Project 3: Home Security System

Enhance your home's safety by integrating a smart security system that utilizes cameras, sensors, and smart locks to monitor and protect your living space.

10.4.1 Objectives

- Monitor key areas of your home in real time.

- Set up automated alerts and notifications for unusual activities.

- Integrate various security devices into a single, cohesive system.

10.4.2 Materials and Tools

- Smart security cameras (indoor and/or outdoor).

- Door and window sensors.

- Smart locks for key entry points.

- A smart hub or a dedicated security app.

- Internet connectivity and a smartphone for remote monitoring.

- Optional: Motion detectors and sirens.

10.4.3 Step-by-Step Guide

1. **Installation:**

   o Mount security cameras in strategic locations to cover entry points and vulnerable areas.

- Install door/window sensors on all accessible openings.

- Replace standard locks with smart locks where possible.

- Ensure all devices are connected to a central hub or directly to your Wi-Fi network.

2. **Configuration:**

- Use the security app to add each device to your system.

- Configure sensor sensitivity and camera motion detection zones.

- Set up smart lock access codes and integrate them with your smartphone.

3. **Automation and Alerts:**

- Create a "Secure Home" routine that locks all doors, arms sensors, and activates cameras when you leave.

- Program automated alerts to notify you via SMS or email if any sensor is triggered.

- Integrate voice assistant commands for quick security checks or to trigger panic alarms.

4. **Testing and Maintenance:**

   o Conduct regular tests of each security component. For instance, trigger door sensors manually to ensure timely alerts.

   o Review recorded footage from cameras to verify the system's responsiveness.

   o Periodically update device firmware to maintain security standards.

10.4.4 Enhancements

- **Integration with Other Systems:**
  Consider linking your security system with your smart lighting or energy management systems—for example, turning on exterior lights when a camera detects motion at night.

- **Remote Monitoring:**
  Use a VPN or secure cloud service to access your security feeds when away from home, ensuring both convenience and data privacy.

10.5 Encouraging Creativity and Customization

The projects provided here are starting points. As you gain confidence, you're encouraged to customize and expand these ideas to suit your unique needs.

10.5.1 Experimentation Ideas

- **Custom Scenes:**
  Develop personalized scenes that combine lighting, sound, and climate control for specific activities like reading, exercise, or entertaining.

- **Advanced Sensor Integration:**
  Experiment with additional sensors (e.g., humidity, CO_2) to further automate your environment and enhance comfort.

- **Third-Party Platforms:**
  Integrate your DIY projects with platforms such as IFTTT, SmartThings, or Home Assistant to create even more sophisticated automations and data visualizations.

10.5.2 Sharing Your Projects

- **Document Your Process:**
  Keep a project journal or blog about your experiences, challenges, and breakthroughs. Sharing your journey can help others and invite feedback for further improvements.

- **Join Communities:**
  Participate in smart home forums, local maker spaces, or online groups where you can exchange ideas, collaborate on projects, and learn about emerging trends.

- **Iterate and Improve:**
  Treat each project as a living work in progress. As new technologies and techniques become available, revisit and refine your installations to enhance performance and functionality.

Final Thoughts

Hands-on projects are an excellent way to bridge the gap between theoretical knowledge and practical application. In Chapter 10, you've learned how to implement smart lighting, energy management, and security systems through detailed, DIY projects. By following these guides and embracing experimentation, you'll not only enhance your smart home's functionality but also develop valuable skills and insights that will serve you well as you continue to innovate.

Remember, every great smart home project starts with a single step. Enjoy the process, be patient with challenges, and let your creativity guide you toward a more intelligent, efficient, and personalized living environment.

Chapter 11: Resources, Glossary, and Further Learning

This chapter is dedicated to providing you with a comprehensive collection of resources that will support your journey in building and managing your smart home. Whether you're looking for detailed technical explanations, community support, or a quick reference guide to the terminology used throughout this book, this chapter is designed to be your go-to companion. It includes curated online and offline resources, a detailed glossary of key terms, and recommendations for further learning that will help you stay updated in the rapidly evolving world of smart home and IoT technology.

11.1 Online Resources and Communities

Staying connected with up-to-date information and active communities is essential for a successful smart home experience. This section lists a variety of websites, forums, and blogs where you can find in-depth tutorials, troubleshooting advice, and the latest news on smart home technologies.

11.1.1 Websites and Blogs

- **IoT For All (iotforall.com):**
  Offers articles, news, and insights into IoT trends, practical tips, and product reviews.

- **CNET's Smart Home Section (cnet.com/topics/smart-home):**
Provides product reviews, buying guides, and the latest news about smart home devices and technology.

- **TechHive (techhive.com):**
Features reviews, comparisons, and how-to guides on a variety of smart home products.

- **The Smart Home Blog (smarthomeblog.net):**
A community-driven site where enthusiasts share DIY projects, automation routines, and tips for troubleshooting.

11.1.2 Forums and Discussion Groups

- **Reddit Communities:**

  - r/homeautomation – A community focused on sharing ideas, projects, and troubleshooting tips.

  - r/IOT – A forum for discussing broader IoT trends and technical challenges.

- **SmartThings Community (community.smartthings.com):**
A dedicated forum for users of Samsung SmartThings and compatible devices, where you can find discussions, custom integrations, and advice.

- **Home Assistant Community Forum (community.home-assistant.io):**
  Ideal for those using the Home Assistant platform, with discussions ranging from beginner questions to advanced configuration.

11.1.3 YouTube Channels and Podcasts

- **YouTube Channels:**

  - "Smart Home Solver" and "Tech With Brett" offer video tutorials, product reviews, and step-by-step guides for various smart home setups.

  - "The Hook Up" provides in-depth discussions and hands-on demonstrations of smart home integrations.

- **Podcasts:**

  - "The IoT Podcast" – Focuses on trends, interviews, and news in the world of IoT.

  - "HomeTech Talk" – Discusses smart home technologies, troubleshooting, and reviews from industry experts.

11.2 Glossary of Key Terms

Understanding the terminology is critical for navigating the complex world of smart home technology. Below is a glossary of commonly used terms and acronyms, explained in plain language.

Common Terms and Definitions

- **Automation:**
  The process of programming devices to perform actions automatically based on predefined triggers, conditions, or schedules.

- **Connectivity Protocols:**
  The set of rules that devices follow to communicate with each other. Common protocols include Wi-Fi, Bluetooth, Zigbee, and Z-Wave.

- **Firmware:**
  The software programmed into a device's hardware that controls its functions. Regular updates help fix bugs and improve performance.

- **Hub/Bridge:**
  A central device that connects multiple smart home devices, enabling them to communicate and be controlled from a single point.

- **IoT (Internet of Things):**
  A network of interconnected devices that can communicate and share data over the internet.

- **Mesh Network:**

  A network topology in which devices (nodes) connect directly, dynamically, and non-hierarchically to as many other nodes as possible, enhancing coverage and reliability.

- **Protocol:**

  A standardized set of rules for data transmission between devices. Examples include HTTP, MQTT, and CoAP in the context of IoT.

- **Routines/Scenes:**

  Predefined sets of actions that are triggered by a single command or event. For instance, a "Goodnight" routine might turn off lights, lock doors, and adjust the thermostat.

- **Smart Home Ecosystem:**

  A collection of devices and services that work together under a unified platform, such as those provided by Amazon Alexa, Google Home, or Apple HomeKit.

- **Voice Assistant:**

  A software agent that uses voice recognition, natural language processing, and speech synthesis to interact with users and control smart devices (e.g., Alexa, Google Assistant, Siri).

- **Zigbee and Z-Wave:**

  Low-power, wireless communication protocols

designed specifically for home automation that enable devices to form mesh networks.

11.3 Further Learning and Continuing Education

To remain at the forefront of smart home technology, continuous learning is key. This section provides suggestions for courses, certifications, books, and online learning platforms that can deepen your knowledge and keep you updated.

11.3.1 Online Courses and Tutorials

- **Coursera and Udemy:**
  Both platforms offer courses on IoT, smart home technology, and home automation. Look for courses that provide hands-on projects and real-world case studies.

- **edX:**
  Offers university-level courses on embedded systems, IoT, and cybersecurity, which can be very useful for understanding the technical underpinnings of smart home devices.

- **YouTube Tutorials:**
  Channels like "The Hook Up" and "Tech With Brett" regularly publish tutorials on integrating and troubleshooting smart home devices.

11.3.2 Books and E-Books

- **"Designing Connected Products: UX for the Consumer Internet of Things" by Claire Rowland et al.:**
  Provides insights into user experience and design principles for IoT devices.

- **"Smart Homes For Dummies" by Danny Briere and Pat Hurley:**
  A beginner-friendly guide that covers the basics of smart home technology and practical installation tips.

- **"Building the Internet of Things" by Maciej Kranz:**
  Offers a comprehensive look at the technical, strategic, and business aspects of IoT, including smart home applications.

11.3.3 Certifications and Professional Development

- **CompTIA IT Fundamentals+ (ITF+):**
  A good starting point for understanding basic IT concepts, which are useful for troubleshooting smart home networks.

- **Cisco's IoT Certification:**
  For those interested in a deeper technical dive, Cisco offers specialized certifications in IoT networking and security.

- **Local Workshops and Maker Spaces:**
  Many communities offer workshops on home automation and IoT. Participating in these events can provide hands-on experience and networking opportunities with other enthusiasts.

Final Thoughts

Chapter 11 is your gateway to a wealth of resources that can support your ongoing journey in smart home technology. By leveraging the websites, communities, glossaries, and further learning opportunities provided here, you'll be well-equipped to tackle new challenges, stay informed about emerging trends, and continuously refine your smart home setup.

Remember, the world of IoT and smart homes is constantly evolving. Keeping a curious mindset and engaging with the broader community will not only enhance your knowledge but also open up new possibilities for innovation in your own smart home.

Chapter 12: Conclusion: Embracing the Future of Smart Living

As we reach the end of this guide, it's time to reflect on the journey you've undertaken to understand, build, and enhance your smart home. This final chapter is dedicated to summarizing the key concepts, reinforcing the transformative potential of smart home technology, and inspiring you to continue exploring new innovations. With technology evolving at a rapid pace, embracing the future of smart living means staying curious, adaptable, and proactive in upgrading your environment.

12.1 Recap of Key Takeaways

Over the course of this guide, you've learned about a wide range of topics that collectively empower you to design and manage a comprehensive smart home system. Here's a brief review of the essential elements:

- **Understanding IoT and Smart Homes:**
  You learned the fundamental concepts behind IoT and how interconnected devices can improve convenience, efficiency, security, and personalization in your daily life.

- **Building Blocks and Connectivity:**
  The guide covered various connectivity protocols (Wi-Fi, Bluetooth, Zigbee, Z-Wave, and emerging standards like Matter), home networking basics,

and the importance of a strong, secure network backbone.

- **Device Selection and Integration:**
  Detailed chapters on choosing devices, setting them up, and integrating them into a cohesive system have equipped you with practical strategies to build an adaptable and future-proof smart home.

- **Automation and Troubleshooting:**
  You explored how to create effective automation routines—from simple schedules to complex multi-device scenarios—as well as systematic approaches for troubleshooting common issues.

- **Security and Future-Proofing:**
  With an emphasis on best practices and practical security measures, you now understand how to protect your devices, data, and network from evolving threats, all while preparing for emerging technologies.

- **Hands-On Projects and Continuous Learning:**
  The DIY projects provided actionable steps to bring smart technology into your everyday life, and the resources, glossary, and further learning sections are designed to support your ongoing journey in this dynamic field.

12.2 The Impact of a Smart Home

Building a smart home isn't just about installing devices; it's about transforming your living space into a dynamic, responsive environment that adapts to your lifestyle. The benefits include:

- **Enhanced Quality of Life:**
  Enjoy greater comfort, convenience, and energy savings as your home anticipates your needs and responds automatically to your routines.

- **Improved Security:**
  With robust security measures, your home becomes a fortress against potential threats, offering peace of mind whether you're at home or away.

- **Sustainable Living:**
  Energy management systems and smart devices work together to optimize consumption, reduce waste, and contribute to a greener planet.

- **Empowered Innovation:**
  The process of building and continually refining your smart home cultivates a mindset of innovation, encouraging you to experiment, learn, and adapt as new technologies emerge.

12.3 Embracing Change and Future Trends

The field of smart home technology is continuously evolving, and staying ahead means being open to change and new ideas. Here are some ways to embrace the future:

- **Stay Informed:**
  Regularly consult the online resources, communities, and further reading materials provided in this guide. Attend workshops, webinars, and industry events to keep your knowledge current.

- **Experiment and Iterate:**
  Treat your smart home as a living project. Regularly experiment with new devices, update automations, and refine your security measures. Innovation often comes from small, iterative changes.

- **Plan for Scalability:**
  As you expand your smart home, maintain a flexible, modular approach. Invest in interoperable devices and scalable platforms that can evolve with emerging technologies, ensuring long-term compatibility and ease of integration.

- **Embrace Emerging Technologies:**
  Keep an eye on trends such as artificial intelligence, machine learning, edge computing, and new communication protocols. These innovations have the potential to further enhance automation,

personalization, and security in your smart home.

12.4 Final Words of Encouragement

Every expert in the realm of smart technology started as a beginner. The journey from understanding basic concepts to implementing sophisticated systems is one of continuous learning and adaptation. As you move forward:

- **Be Patient and Persistent:**
  Building a smart home is an ongoing process. Embrace the challenges as opportunities to learn and grow.

- **Share Your Knowledge:**
  Engage with communities, help others troubleshoot issues, and share your own success stories. The collective wisdom of the smart home community is one of its greatest assets.

- **Keep Innovating:**
  Your smart home is not the endpoint—it's a foundation upon which you can build ever more intelligent, efficient, and secure living spaces. Let your curiosity drive you to explore and incorporate new technologies as they become available.

Final Thoughts

As we conclude this guide, remember that smart home technology is more than a collection of gadgets—it's a pathway to a more connected, efficient, and secure lifestyle. By embracing the principles and strategies outlined in this book, you are not only transforming your home but also positioning yourself at the forefront of a technological revolution.

Thank you for taking the time to explore the world of smart home and IoT for beginners. May your journey toward a smarter, more integrated home be as exciting and rewarding as the destination itself. The future of smart living is in your hands—embrace it with confidence, creativity, and a commitment to continuous learning.

Appendix A: Quick Reference Guides, Checklists, and Flowcharts

This appendix is designed to serve as a handy resource for quickly reviewing key tasks, troubleshooting common issues, and planning your smart home projects. It contains several checklists, flowcharts, and templates that you can use to streamline your smart home setup, maintenance, and automation processes. Whether you're a beginner or a seasoned enthusiast, these quick reference tools will help you stay organized and efficient as you work with your smart home devices.

A.1 Smart Home Setup Checklist

Use this checklist as a step-by-step guide when installing new devices or performing major upgrades to your smart home system.

Pre-Installation

- **Inventory Verification:**
  - Confirm that all items (devices, cables, adapters, mounting hardware, manuals) are included in the package.
  - Inspect devices for any physical damage.

- **Review Documentation:**
  - Read the quick-start guide and

manufacturer's instructions.

- o Note any specific requirements (e.g., minimum Wi-Fi signal strength, power needs, mounting considerations).

- **Prepare Your Environment:**

  - o Identify optimal installation locations based on signal strength, accessibility, and usage.

  - o Ensure your home network is secure and functioning properly.

Installation Steps

- **Unboxing and Physical Setup:**

  - o Carefully unpack and inspect each device.

  - o Install batteries or connect to a power source as needed.

  - o Mount or place devices according to manufacturer guidelines.

- **Network and Connectivity:**

  - o Ensure that devices are within range of your Wi-Fi router or hub.

  - o Follow on-screen instructions to connect the device to your home network.

  - o Verify connectivity by checking the device status in the companion app.

- **Firmware and Software Updates:**

  - Check for and install any available firmware updates immediately after initial setup.

  - Confirm that the device's companion app is updated to the latest version.

- **Initial Testing:**

  - Test the basic functionality of each device (e.g., turning on/off, responding to commands).

  - Verify that any automation routines or integrations (if pre-configured) work as expected.

A.2 Troubleshooting Flowcharts

Use these flowcharts as a visual guide to diagnose and resolve common issues that may arise in your smart home.

A.2.1 Connectivity Issues Flowchart

1. **Issue: Device Not Responding**
   ↓

2. **Check Power Source:**

   - Is the device powered on?

     - **Yes:** → Proceed to step 3.

- **No:** → Reconnect or replace batteries; then test again. ↓

3. **Verify Network Connection:**

   o Is the device connected to your Wi-Fi network?

      - **Yes:** → Proceed to step 4.

      - **No:** → Re-run the network setup process. ↓

4. **Check Signal Strength:**

   o Is the device within optimal range of your router or hub?

      - **Yes:** → Proceed to step 5.

      - **No:** → Relocate the device or extend network coverage (e.g., use a mesh network). ↓

5. **Update Firmware:**

   o Are firmware and software versions up-to-date?

      - **Yes:** → If issues persist, consult the manufacturer's support.

      - **No:** → Update firmware and test again.

A.2.2 Device Pairing Issues Flowchart

1. **Issue: Device Not Pairing**

   ↓

2. **Confirm Pairing Mode:**

   o Is the device in pairing mode (check LED indicators or manual instructions)?

      - **Yes:** → Proceed to step 3.

      - **No:** → Activate pairing mode according to the manual. ↓

3. **Verify Proximity:**

   o Is the device close enough to the hub or smartphone?

      - **Yes:** → Proceed to step 4.

      - **No:** → Bring the device closer and try pairing again. ↓

4. **Check for Interference:**

   o Are there nearby devices causing interference (e.g., other wireless devices, microwaves)?

      - **Yes:** → Minimize interference and try again.

      - **No:** → Proceed to step 5. ↓

5. **Update and Restart:**

   ○ Restart the device, hub, or smartphone.

   ○ Update firmware if available.

   ○ Attempt pairing again. ↓

6. **If Still Failing:**

   ○ Consult troubleshooting documentation or contact support.

A.3 Routine Maintenance and Update Schedules

Establish a regular maintenance routine to keep your smart home system running smoothly and securely.

Weekly Tasks

- **Check Device Status:**

   ○ Open companion apps and verify that all devices are online.

   ○ Look for any alerts or error messages.

- **Network Performance:**

   ○ Run a quick network diagnostic (using built-in app features or a Wi-Fi analyzer) to ensure optimal connectivity.

Monthly Tasks

- **Firmware and Software Updates:**

  - Check for updates for all devices, routers, hubs, and apps.

  - Install updates and restart devices if needed.

- **Review Automations:**

  - Test critical automation routines to ensure they trigger as expected.

  - Update routines if new devices have been added or if usage patterns have changed.

Quarterly Tasks

- **Full System Audit:**

  - Revisit the device inventory and update documentation.

  - Assess the performance of the entire smart home system and note any recurring issues.

  - Check security settings, change passwords if necessary, and verify network segmentation (e.g., guest network isolation).

- **Backup Configurations:**

  - Export and save configuration settings and automation routines from companion apps, where possible.

- o Document any custom integrations or scripts used.

A.4 Automation Planning Template

Use this template to plan and document your automation routines. Customize the template based on your specific needs.

Automation Routine Template

- **Routine Name:**
  (e.g., "Morning Wake-Up Routine")

- **Objective:**
  (Describe what the routine is intended to accomplish, such as gradually brightening lights, adjusting the thermostat, and playing a morning playlist.)

- **Trigger(s):**
  (List the event(s) that will activate the routine, such as a specific time, motion detection, or voice command.)

- **Action(s):**
  (Detail the sequence of actions that will occur. Include device names, specific commands, and order of execution.)

○ Example:

1. Gradually increase bedroom light brightness over 5 minutes.

2. Set thermostat to 72°F.

3. Play selected music on the smart speaker.

- **Conditions:**
(Specify any conditions that must be met for the routine to execute, such as only on weekdays, or only when the ambient light is below a certain level.)

- **Testing and Verification:**
(Describe how you will test the routine, note any modifications made, and document the final setup for future reference.)

- **Notes:**
(Any additional comments, potential improvements, or integration ideas.)

Final Thoughts on Appendix A

This appendix is intended to be a living document that you update as your smart home system evolves. Whether you're troubleshooting issues, planning new automations, or scheduling regular maintenance, these checklists, flowcharts, and templates are here to ensure that your smart home remains efficient, secure, and responsive to your needs.

Keep this appendix handy and refer back to it as a quick reference guide whenever you need to perform routine tasks or solve problems. As you grow more comfortable with your smart home system, consider refining these tools to better suit your specific setup and workflow.

Appendix B: Technical Resources, Device Specifications, and Advanced Integrations

This appendix is designed for those who want to delve deeper into the technical aspects of smart home systems. It provides detailed device specifications, sample integration code, advanced troubleshooting techniques, and links to vendor and developer resources. Whether you're a tech enthusiast seeking to customize your smart home or a professional aiming to integrate advanced functionalities, this appendix serves as a comprehensive technical resource.

B.1 Detailed Device Specifications and Datasheets

Understanding the technical specifications of your smart home devices is critical for optimal integration and troubleshooting. This section provides guidance on interpreting datasheets, technical documents, and common parameters for typical smart home components.

B.1.1 Common Device Parameters

- **Connectivity Standards:**

  - **Wi-Fi:** Frequency bands (2.4 GHz, 5 GHz), data rates, range, and encryption protocols (WPA2/WPA3).

  - **Bluetooth:** Versions (Classic, BLE), range, data throughput, and pairing mechanisms.

- **Zigbee/Z-Wave:** Mesh networking capabilities, channel frequencies, data rates, and power consumption.

- **Power Requirements:**

  - **Voltage and Current:** Operating voltage, power consumption (in watts or mA), and energy-saving modes.

  - **Battery Life:** For battery-operated devices, typical life span under standard operating conditions and standby modes.

- **Sensor Specifications:**

  - **Accuracy and Precision:** Measurement tolerances for temperature, motion, light, or other sensor types.

  - **Response Time:** How quickly the sensor detects changes and communicates data.

- **Firmware and Software Capabilities:**

  - **Update Mechanisms:** Over-the-air (OTA) update capabilities, versioning, and compatibility with companion apps.

  - **APIs and Integration Support:** Availability of open APIs, SDKs, or developer documentation for third-party integrations.

B.1.2 Sample Datasheet Template

Below is a simplified template that manufacturers often follow in their datasheets. Use this template as a checklist when evaluating a device's technical documentation:

Parameter	Specification	Notes
Connectivity	Wi-Fi (2.4 GHz/5 GHz)	Supports WPA3 encryption
Power Consumption	5 W average (standby: 0.5 W)	Typical for smart bulbs
Operating Voltage	100-240 V AC	Compatible with global standards
Sensor Accuracy	±0.5°C (temperature sensors)	Ensures precise climate control
Firmware Update	OTA supported	Automatic and manual update options
API Availability	RESTful API, MQTT support	Developer documentation available

B.2 Integration Code Samples and API References

This section offers sample code snippets and guidance for integrating smart home devices with popular automation platforms. The examples provided are designed to work with open platforms like IFTTT, Home Assistant, or custom integrations using RESTful APIs.

B.2.1 Sample Code: Triggering an Automation via IFTTT

Below is an example of a simple IFTTT applet using a webhook to trigger a smart home routine:

- json
- Copy
- {
-   "ifttt_event": "motion_detected",
-   "value1": "Living Room Motion Sensor",
-   "value2": "Trigger Time: {{OccurredAt}}",
-   "value3": "Initiate Night Mode"
- }
- **Explanation:**
  - The JSON payload is sent when a motion sensor detects movement.
  - ifttt_event identifies the event type.

- value1, value2, and value3 carry additional data used in the automation routine.

B.2.2 Sample Code: Home Assistant YAML Configuration

Home Assistant is a popular open-source platform that supports YAML configuration files for creating custom automations. The following snippet automates turning on a light when motion is detected:

- yaml

- Copy

- automation:

- - alias: "Turn on Living Room Light"

- trigger:

- - platform: state

- entity_id: binary_sensor.living_room_motion

- to: 'on'

- condition:

- - condition: time

- after: "18:00:00"

- before: "23:00:00"

- action:

- - service: light.turn_on

- entity_id: light.living_room

- **Explanation:**

  - The automation triggers when the living room motion sensor turns on.

  - A time condition restricts the automation to evening hours.

  - The action calls a service to turn on the living room light.

B.2.3 API References and Developer Portals

- **IFTTT Developer Documentation:**
  Visit IFTTT Platform for detailed guides on creating custom applets and integrating webhooks.

- **Home Assistant Developer Docs:**
  Access the Home Assistant Developer Documentation for tutorials, API references, and custom component development.

- **Vendor APIs:**
  Many smart home devices offer APIs. Check manufacturer websites (e.g., Philips Hue, Nest, Samsung SmartThings) for API documentation and integration examples.

B.3 Advanced Troubleshooting and Diagnostic Tools

For those encountering complex issues or seeking deeper insights into system performance, this section provides advanced troubleshooting techniques and tool recommendations.

B.3.1 Network Diagnostic Tools

- **Wi-Fi Analyzers:**
  Tools like NetSpot, WiFi Analyzer (Android), or AirPort Utility (iOS) can help measure signal strength, interference, and channel congestion.

- **Packet Sniffers:**
  Use tools like Wireshark to capture and analyze network packets. This can be invaluable for diagnosing communication issues between devices.

B.3.2 Logging and Monitoring Techniques

- **Device Logs:**
  Many companion apps provide access to device logs that detail connection events, errors, and firmware updates. Regularly review these logs for patterns or recurring issues.

- **Centralized Monitoring:**
  Platforms like Home Assistant offer dashboards that aggregate device status, network performance, and automation logs. Custom scripts can also be developed to send alerts based on log events.

B.3.3 Advanced Troubleshooting Checklist

- **Step 1:** Identify the issue by reviewing error messages and logs.

- **Step 2:** Verify network connectivity and device power supply.

- **Step 3:** Use diagnostic tools (e.g., Wireshark) to analyze data traffic.

- **Step 4:** Cross-reference device firmware and update logs.

- **Step 5:** Consult vendor technical support with detailed logs and diagnostic data.

B.4 Vendor and Manufacturer Technical Resources

Staying informed about device updates, firmware releases, and integration improvements is key to maintaining a robust smart home system. Below is a list of valuable vendor and manufacturer resources:

- **Philips Hue Developer Portal:**
  Hue Developer provides APIs, SDKs, and detailed technical documentation for smart lighting products.

- **Nest Developer Documentation:**
  Access the Nest Developer Center for integration guides on smart thermostats and security cameras.

- **Samsung SmartThings Developer Portal:**
  Visit SmartThings Developer for comprehensive guides on building and integrating with the SmartThings ecosystem.

- **Apple HomeKit:**
  The HomeKit Developer page offers technical documentation, best practices, and code samples for integrating HomeKit-compatible devices.

B.5 Emerging Trends and Future Technologies

The smart home landscape is evolving rapidly with new technologies and standards. This section highlights emerging trends that may influence future integrations.

B.5.1 Artificial Intelligence and Machine Learning

- **Adaptive Automation:**
  AI can enhance automation by learning user behaviors and predicting needs, leading to more intelligent and responsive systems.

- **Predictive Maintenance:**
  Machine learning algorithms can analyze sensor data to predict device failures and optimize maintenance schedules.

B.5.2 Edge Computing

- **Local Processing:**
  Shifting data processing from the cloud to local devices reduces latency, enhances privacy, and improves reliability during network outages.

- **Decentralized Systems:**
  Edge computing supports the development of decentralized smart home networks, making systems more resilient and scalable.

B.5.3 Universal Standards and Interoperability

- **Matter Protocol:**
  The upcoming Matter standard promises improved interoperability across devices from different manufacturers, simplifying integration and reducing vendor lock-in.

- **Thread Networking:**
  A low-power mesh networking protocol, Thread is designed for IoT devices and offers enhanced security and reliability.

B.5.4 Future-Proofing Strategies

- **Modular Systems:**
  Invest in devices and hubs that support modular upgrades and are designed to integrate with emerging standards.

- **Continuous Learning:**
  Stay engaged with industry developments through webinars, technical conferences, and online communities dedicated to IoT and smart home technologies.

Final Thoughts on Appendix B

Appendix B is intended to be a dynamic resource that evolves with your smart home system and the broader IoT landscape. By understanding detailed device specifications, leveraging integration code samples, employing advanced troubleshooting tools, and staying informed about emerging technologies, you can significantly enhance the functionality and resilience of your smart home.

Keep this appendix handy as you work on advanced integrations, customize your setup, or troubleshoot complex issues. As new technologies emerge, revisit and update these resources to ensure your smart home remains at the forefront of innovation.

www.ingramcontent.com/pod-product-compliance
Lightning Source LLC
LaVergne TN
LVHW021459170326
834004LV00004B/351